BUILT TO SU[...]

BIRDS OF PREY

Alan Walker

TABLE OF CONTENTS

A Crabtree Seedlings Book

CRABTREE
Publishing Company
www.crabtreebooks.com

peregrine
falcon

Birds of Prey

Birds of prey catch and eat other animals.

A peregrine falcon is the fastest animal in the world. It can dive through the air at speeds up to 200 mph (322 kph).

Birds of prey are called raptors.

white-tailed sea eagles

Owls, hawks, eagles, vultures, kites, and falcons are raptors.

golden eagle

Raptors have talons and sharp, hooked beaks.

The golden eagle is North America's largest bird of prey.

Raptors use their strong feet and sharp talons to grab prey.

harpy eagle

Harpy eagle talons can be as large as grizzly bear claws!

red-tailed hawk

They use their sharp beaks to tear through **flesh**.

Red-tailed hawks are the most common hawks in North America.

Raptors have excellent eyesight.

An eagle can be hundreds of feet in the air and see a fish in the water below.

bald eagle

Most owls are **nocturnal** and have excellent night vision.

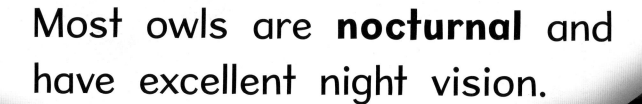

Owls cannot turn their eyes. They turn their heads to see in different directions.

boreal owl

15

barn owl

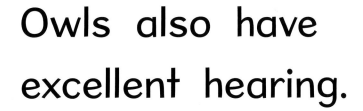

Owls also have excellent hearing.

A barn owl can hear the tiny footsteps of a mouse on the soft forest floor— even while flying!

Vultures do not hunt and kill their food.

griffon vultures

Vultures search for and eat **carrion**.

Vultures use their sense of smell to find carrion.

Raptors live on every **continent** except Antarctica.

North America

The northern crested caracara is Mexico's national bird.

South America

A vulture called the Andean condor is one of the largest flying birds in the world.

Common kestrels live throughout most of Europe.

Phillipine eagles hunt and eat monkeys in the Phillipine rainforests.

Long-legged secretary birds are raptors that live in Africa.

Europe

Asia

Africa

Australia

Antarctica

Glossary

carrion (KAIR-ee-uhn): Carrion is the rotting flesh of a dead animal.

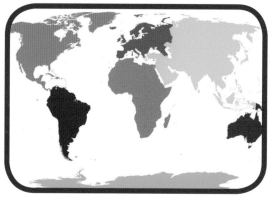

continent (KON-tuh-nuhnt): A continent is one of Earth's seven land masses. Asia, Africa, Australia, Antarctica, Europe, North America, and South America are Earth's continents.

flesh (FLESH): Flesh is the meat of an animal.

nocturnal (noc-TUR-nuhl): Nocturnal animals are active at night. Owls are nocturnal.

Index

The raptor on the front cover is a golden eagle.

School-to-Home Support for Caregivers and Teachers

This book helps children grow by letting them practice reading. Here are a few guiding questions to help the reader build his or her comprehension skills. Possible answers appear here in red.

Before Reading

- **What do I think this book is about?** *I think this book is about birds of prey. I think this book is about birds that eat small animals.*

- **What do I want to learn about this topic?** *I want to learn how fast birds of prey can fly. I want to learn how birds of prey can see small animals while they are flying high in the sky.*

During Reading

- **I wonder why...** *I wonder why vultures don't hunt for food. I wonder why and how vultures can find dead animals.*

- **What have I learned so far?** *I have learned that vultures use their sense of smell to find dead animals. I have learned that owls have excellent hearing and can hear the tiny footsteps of a mouse on the ground.*

After Reading

- **What details did I learn about this topic?** *I have learned that a peregrine falcon can dive through the air at 200 mph (322 kph). I have learned that birds of prey are called raptors and they have talons and sharp, hooked beaks.*

- **Read the book again and look for the glossary words.** *I see the word* **nocturnal** *on page 14, and the word* **carrion** *on page 19. The other glossary words are found on page 22.*

Library and Archives Canada Cataloguing in Publication

CIP available at Library and Archives Canada

Library of Congress Cataloging-in-Publication Data

CIP available at Library of Congress

Crabtree Publishing Company
www.crabtreebooks.com   1–800–387–7650

Written by Alan Walker

Print coordinator: Katherine Berti

Printed in the U.S.A./062021/CG20210401

Print book version produced jointly with Blue Door Education in 2022

Photo credits:
Cover © martellostudio; page 2-3 full page phot © Collins93, falcon inset © Smiler99; page 4-5 © ArCaLu; page 6-7 © Ian Duffield; page 8-9 Harpy Eagle © Chepe Nicoli; page 10-11 © yhelfman; page 12-13 © FloridaStock; page 14-15 © Stanislav Duben; page 16-17 © sirtravelalot; page 18-19 © Valerijs Novickis; page 20-21 map © Peter Hermes Furian, crested caracara © Chepe Nicoli, Andean condor © aabeele, common kestrel © Maria Gaellman, Philippine eagle © Casper Simon, secretary bird © Dmussman All photos from Shutterstock.com

Published in the United States
Crabtree Publishing
347 Fifth Ave.
Suite 1402-145
New York, NY 10016

Published in Canada
Crabtree Publishing
616 Welland Ave.
St. Catharines, Ontario
L2M 5V6